Gohar Marikyan, Ph.D.

ANANIA SHIRAKATSI'S

TVABANUTIUN: WORLD'S OLDEST

MANUSCRIPT ON ARITHMETIC

PART 1: ADDITION

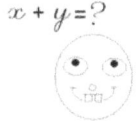

New York, U.S.A.

2011

Anania Shirakatsi's Tvabanutiun: World's Oldest Manuscript
On Arithmetic
Part 1: Addition

by Gohar Marikyan, Ph.D.

Anania Shirakatsi's 7th century manuscript *Tvabanutiun* (Arithmetic) is
the world's oldest preserved manuscript on arithmetic. It contains
interesting detailed explanations of the methods he has developed for
teaching arithmetic to beginners and to first grade children. Shirakatsi's
methods have been successfully used over centuries in Armenia. Even
though times have changed, his methodology continues to be as effective
in the contemporary diverse classroom environment as it has been since
the 7th century.

New York, 2011

ISBN-13: 978-1463761783
ISBN-10: 1463761783

To my father
Gerasim Marikyan, Ph.D.

"I love Armenian people - all of them. I love them because they are a part of the enormous human race, which of course I find simultaneously beautiful and vulnerable."

William Saroyan

Content

Preface

Anania Shirakatsi, in his 7th century manuscript Tvabanutiun (Arithmetic) describes the methods he has developed for teaching arithmetic to beginners and to first grade children. This manuscript is the world's oldest preserved complete manuscript on arithmetic (Hewsen 1968, pg.42). It contains interesting detailed explanations. Shirakatsi's methods have been successfully used over centuries in Armenia. The net effect of his teachings has been the acquisition of higher levels of arithmetic knowledge by children, setting a strong foundation for achievements in scientific research in their adult years. Even though times have changed his methodology continues to be as effective in the contemporary diverse classroom environment as it has been since the 7th century.

This book is only a synopsis and a brief introduction to Anania Shirakatsi's genius and methodology in teaching arithmetic. There is much more to research and to talk about this fascinating topic and about the beauty of his work. The aim of this book is to reflect on his methodology for teaching the addition to beginners and to children.

1. Historical Background

1.1. Introduction

Mathematics, as a science, is one of the pillars of civilization. Even though history may ascribe its birth to a particular nation, the fact is that mathematics, most probably, started with prehistoric man just as soon as he could think, and most probably its origin is connected to the ten fingers that we have. Perhaps that is the reason why we use the ten-based system albeit there have been civilizations that have used other bases for counting. When it comes to teaching mathematics every nation has its own unique method that has been developed over their own history. A few of these methods are interesting indeed. The study of history of mathematics helps the contemporary educator determine the necessary tools for teaching mathematics to today's students.

Quite a number of old manuscripts contain a wealth of information that shed light to the historic development of mathematics or nurture the continuation of current thinking. When it comes to major contributions, one such nation is Armenia, and one particular person in that nation is the 7th century mathematician Anania Shirakatsi, the author of the world's oldest partially preserved manuscript on arithmetic.

In this book I will review and reflect on Anania Shirakatsi's fascinating methodology and his ingenious approach to teaching arithmetic to beginners and to children.

Although Anania Shirakatsi has left a vast array of scientific works, their discussion is outside the scope of this book. I will briefly discuss the history of the manuscript and its author. Additional information about this seventh century Armenian mathematician, writer, historian, geographer, cosmologist, and philosopher can be found in

Hewsen's article   (Hewsen 1968, includes excellent

footnotes and references), Abrahamyan's book (Abraha-

**Picture 1. Statue of Anania Shirakatsi at Matenadaran,
Yerevan, Armenia.**

myan 1944), Hacikyan's book (Hacikyan 2002), and others.

It should be noted that quite often people use the word mathematics instead of the word arithmetic even though the latter is only one part of the former. Briefly, the word mathematics, going back to the 15th century, is from the Greek word mathēma, learning. While arithmetic (from Greek arithmein, to count, from arithmos, number[1]) is a branch of mathematics that deals with addition, subtraction, multiplication, and division of non-negative numbers. I refer here to very basic arithmetic, that is, the very fundamental art or science of adding and subtracting numbers.

---

[1] Merriam Webster's Collegiate Dictionary, 12th Edition

1.2. Anania Shirakatsi's Manuscript

At the end of the 4th Century the necessity emerged to revive the lost Armenian alphabet. King Vramshapouh and

Picture 2: Statue of Mesrop Mashtots with the Armenian alphabet, at Matenadaran, Yerevan, Armenia.

Catholicos Sahak Partev assigned the task to Mesrop Mashtots, a genius scholar monk. In 405, Saint Mesrop (see Picture 2), also known as Mesrop Mashtots, re-created the

13

alphabet for the Armenian language, consisting of 36 letters. To date it remains unchanged, and it is currently in use by Armenians (later, three more letters were added). In a short period of time Mashtots's 36 letters were put to use in translating the Bible, and it precipitated translations from other languages and it encouraged writing in various fields of science, history, and literature. For the next two centuries, political unrest paralleled the exceptional development of literary and religious life that became known as the first Golden Age of Armenian literature (5th to 7th centuries). The 36 letters of the alphabet also replaced the previously used numbering system and it was Armenia's official system over the next 1200 years. It has survived the adoption of the internationally used Hindu-Arabic numerals and it continues to be used in Armenian religious literature and other writings.

My hero is Anania Shirakatsi, the 7th century Armenian mathematician (see Picture 1). His manuscripts have been of great interest to scholars over centuries. Grigor Magistros

Picture 3: Anania Shirakatsi's manuscript displayed at the Matenadaran in Yerevan, Armenia.

(990-1058), an Armenian linguist, scholar and public functionary, among other manuscripts of scientific or philosophical value also owned several works of Anania Shirakatsi. He states that Shirakatsi's works were of great interest to scholars of the time[2]. Today, there remains only a small portion of his manuscripts that has been handed down to us. They are currently among a very vast collection of

---

[2] http://www.nationmaster.com/encyclopedia/Grigor-Magistros

manuscripts safeguarded at the Matenadaran (see Picture 4),

Picture 4: Matenadaran, Yerevan, Armenia.

Armenia's state repository of ancient manuscripts. The Matenadaran, carved in a mountainside, in the capital of Armenia, Yerevan, reportedly contains some 17,000 irreplaceable manuscripts in several ancient languages (among them Greek, Latin, Arabic, Fari/Iranian, Assyrian, Hebrew, Ethiopian, Indian), and in total more than 300,000 documents, thus, the largest collection in the world. Picture 3 is a photograph of Shirakatsi's Tvabanutiun manuscript displayed at the Matenadaran.

ANANIA SHIRAKATSI'S TVABANUTIUN

Shirakatsi's survived manuscripts are: *Inqnakensagrutiun* (Autobiography), *Tvabanutiun* (Arithmetic), *Tiezeragitutiun* (Cosmology), *Tomaragitutiun* (Chronicon[3]), *Ashkharhatsuyts* (Geography), *Odyerevuitabanutiun* (Meteorology), Chapagitutiun (Weights and Measures).

Anania Shirakatsi is one of a few Armenian authors that left a detailed autobiography. It contains not only information about his life but also information about the level of sciences at that time.

Anania's year of birth is generally known to be between 595 and 600 A.D. His name, as was the custom in the olden days simply means Anania of Shirak, Shirak being a province in Armenia.

Shirakatsi was a well-educated young man. He studied Armenian literature and the Holly Scriptures, and was highly impressed by references to wisdom in the Holy Bible ("The wise shall inherit glory, but shame shall be the legacy of

---

[3] Comparative chronology of celestial events

fools." Prov. 3:35, and the like). Anania developed great desire and passion for knowledge. In his own words, "With great admiration of the art of calculation, I came to the conclusion that without numbers nothing can be reasoned, therefore, I considered it the mother of all knowledge." (Shirakatsi, Autobiography 7th Century)

This approach and the love for mathematics and sciences has been passed down through generations and I, a child of the 20th century was a recipient of Shirakatsi's message, and I was taught to love and to respect mathematics, and knowledge (education) in general. Isn't that amazing, to be a recipient of a 7th century message?

In search of a teacher, on the recommendation of his friends, Anania went to Trebizond to meet Tychikos, a famous scholar and professor[4], who accepted him as his pupil. Shirakatsi studied under Tychikos for eight years. After returning home in 651 Anania immediately founded his

---

[4] Rosenqvist 2005, p. 33

own school. The 7th century manuscript *Tvabanutiun* (Arithmetic) is Anania Shirakatsi's textbook that he used for teaching arithmetic. In it he describes methods he developed and used in teaching arithmetic to beginners and to first grade children.

Maintaining its commanding aura, as I can best translate from the Armenian text, Shirakatsi writes:

"You, who seek wisdom and wish to study with me, as a good teacher my aim is to write for you the science of computing that has been developed by the efforts of our ancestors. Learn the tables that I have formulated. Although I have abridged the existing comprehensive version to avoid boring you with lengthy repetitions, I have also simplified the original concept so that you understand it profoundly and thoroughly. So, I start with the most fundamental and the simplest, having in mind the learning level of children and the uneducated."[5]

---

[5] Aghayan 1979, p.30

Shirakatsi's emphasis being on children and the novice, he avoids teaching by rote and prefers teaching by understanding. If the child understands why and how do two numbers add together, he does not have to memorize anything, any step, any method, any formula, or any algorithm. This "understanding" begins with the most fundamental of the four operations of arithmetic, the addition.

"Nakhavarjzoom, the first chapter of arithmetic, is called Addition."[6]

By "the most fundamental and the simplest" Anania Shirakatsi refers to the operation of addition, and he refers to its process of learning as *nakhavarjzoom* (preparatory). A good understanding of addition makes a strong foundation for learning the other three operations, and facilitates the learning process in acquiring further knowledge.

The appendix of this manuscript contains several "Prob-

---

[6] Aghayan 1979, p.30

lems of Amusement" that Shirakatsi had meant for mathematical entertainment in social gatherings. Scholars that have studied this manuscript are of the opinion that perhaps it contained a theoretical part that has not survived.

"Shirakatsi's role in spreading mathematical knowledge in Armenia is especially significant if one realizes that the period between A.D. 500 and 800 is generally considered the dark ages of mathematics in the West. The Greek Proclus (c. 410-85) and the Roman Boethius (c. 480-524) were the last glowing embers of the dying fire of mathematics; in fact, the historian Edward Gibbon characterizes them as 'the last of the Romans whom Cato or Tully could have acknowledged for their countrymen.' (Quoted by D. E. Smith in History of Mathematics, New York, 1958, p. 102.) Some three hundred years were to pass before mathematics was revived by Arabs, in the ninth century."[7]

---

[7] Hacikyan 2002, p. 57

1.3. Numbering Systems

In ancient times, before Hindu-Arabic numerals were invented, Greeks, Armenians, and many other nations used letters for numbers. This system is based on assigning values to the letters of the alphabet. Just after re-creation of the Armenian alphabet in 405 A.D. by Mesrop Mashtots the 36 letters replaced the previously used numbering system in Armenia. Mashtots arranged his letters of the alphabet in four columns with nine letters in each column. This system consists of arranging the 36 letters into four columns, each column progressing in a vertical fashion top to bottom, and then continuing to the next column to the right. Most depictions of the Armenian alphabet are in this fashion (See Picture 1 and Picture 5). The first column represents ones, the second – tens, then comes hundreds, and the fourth column represents thousands. Therefore, each column represents one place value. There was no letter and/or notation for zero in this system. In fact, because this system was not a positional system, there was no need to assign a symbol to zero. As a result, numbers in this system are represented in

Ones		Tens		Hundreds		Thousands	
Letter	Numeric Value	Letter	Numeric Value	Letter	Numeric Value	Letter	Numeric Value
Ա	1	Ժ	10	Ճ	100	Ռ	1000
Բ	2	Ի	20	Մ	200	Ս	2000
Գ	3	Լ	30	Յ	300	Վ	3000
Դ	4	Խ	40	Ն	400	Տ	4000
Ե	5	Ծ	50	Շ	500	Ր	5000
Զ	6	Կ	60	Ո	600	Ց	6000
Է	7	Հ	70	Չ	700	Ւ	7000
Ը	8	Ձ	80	Պ	800	Փ	8000
Թ	9	Ղ	90	Ջ	900	Ք	9000

Picture 5: Armenian alphabet with corresponding numeric values

more efficient way. For example, the number 3000 is represented by one letter (see Picture 5). In order to write a number, the numeric values for individual letters would be added. The letter with a higher place value is placed to the left of those with lower place values. For example, to write 2009 we need to write the letter with numeric value two thousand followed by the letter corresponding to nine (see Picture 5, p. 23). Below is 2009 in Armenian alphabetical numerals.

Ս Թ

Thus, the numeric values for individual letters should be added to get the total numeral value. Correspondingly,

Թ Ջ Թ Թ

represents 9999. Additional signs in conjunction with letters were used to indicate numbers higher than 9999. The sign *byur*, similar to ∧ above a numeral was used to indicate that the numeral is multiplied by 10,000 (see Table A). The principle behind this system is similar to the Ancient Greek numeral system. However, there were major differences in

Table A: Shirakatsi's tables of addition.

indicating larger numbers. Using only the 27 letters in the ancient Greek alphabet, the highest possible number is 999. Additional signs had to be used for higher numbers. The shape and the position of the additional sign is another difference in these two systems. Likewise, there are similarities and differences in indicating fractions in the two systems, however, this discussion is outside the scope of this book.

2. Anania Shirakatsi's Methodology

2.1. Net Effect of Anania Shirakatsi's Methodology

My research is based on only one part of the survived manuscript *Tvabanutiun* (Arithmetic), addition, which relates to the very fundamental part and the basic structure of his methodology. It contains interesting, detailed explanations of his methods that have been used for centuries in Armenia. The net effect of his teachings has resulted in the acquisition of a high degree of arithmetic knowledge by children and a strong foundation for subsequent achievements in scientific research in their mature age. As I will show, and explain, I am of the opinion that Anania Shirakatsi's teaching methodology can be as effective in the contemporary diverse classroom as it has been since the 7th century.

2.2. Anania Shirakatsi's Tables of Addition

In this book I limit my discussion only to addition. In ancient Armenian literature addition was referred to as *endunelutiun*, (receive) in the sense that one number receives another number to complete the addition.

Shirakatsi's tables of addition consist of four groups with nine tables in each. In the manuscript all four groups are placed on a single page. Each group is placed in one column (see Table A, p. 25). Please note that the table does not include the symbols + and =. The symbol + indicating addition was used for the first time in the 15th century, and the symbol = indicating equality was first used in the 16th century[8]. To show the concept more clearly I have also shown Shirakatsi's tables in Hindu-Arabic numerals instead of Armenian letters (see Table B). Notice that in Table A the first addend in each tablet is the same and it progresses downward in the alphabetical order. The alphabetical order eases memorization. (We can check the same in Table B, p. 32. The first addend in each tablet repeats, and it progresses downward from 1 to 9 in the first column, from 10 to 90 in the second column, etc.) I will expand more on the significance of this setup after I describe the first group of tablets.

The first group is shown in the first column of the table. It shows the addition of single digit numbers. Please note that in this table 2+1=3 is missing, as are 3+2, 4+3, etc., and all columns start with the addition of the number and itself.

Why so? What can this table teach other than addition? It teaches that addition is augmentation, and that the order of

---

[8] Ball 1901, p. 201

addends is unimportant, that is, it teaches the commutative property of addition, and how to use it. Is it essential for a child to learn it? Absolutely! With this method the child understands the meaning of addition. 1+2 is the same as 2+1. It shows the child that addition is about putting two quantities together, regardless of order. Therefore, the student understands that adding 2 and 1 is the same as adding 1 and 2. The commutative property of addition is a very important discovery and a source of knowledge for the child. It becomes second nature, and it makes lasting impression in a child's mind. It helps to perform additions faster and intuitively.

2.2.1. First Column: the Addition of Single Digit Numbers

The first tablet teaches the addition of 1 to numbers 1 through 9. It shows how numbers grow from 1 to 10. 1+1 is 2, 1+2 is 3, etc. The child understands that to count is the same as to add 1 to the preceding number. The child learns that 1+3 is 4, and 1+4 is 5, that is, twice adding 1 to 3 we get 5, therefore, 5 is greater than 3, that 5 is 2 ones away from 3, and that the difference of 3 and 5 is 2. This also creates a good foundation for understanding subtraction. These are important discoveries for the child. The knowledge gained by these discoveries forever stay with the child.

Tablet two shows the addition of 2 to numbers 2 through 9. This tablet is one line shorter than the previous one. Tablet three shows the addition of 3 to numbers 3 through 9. Notice that this tablet, too, is one line shorter than the previous tablet. Each following tablet is shorter by one line. This, also, makes the child to stop and think. Finally, the last tablet shows the addition of 9 to 9, and consists of only one line. The student, looking at the tablets of the first column will notice that the total of any two numbers in one tablet can also be found in other tablets. For example, three consecutive tablets show 1+5=6, 2+4=6 and 3+3=6. Thus, the student will

conclude that 6 is the addition of 1 and 5, or 2 and 4, or 3 and

1 + 1 = 2	10 + 10 = 20	100 + 100 = 200	1000 + 1000 = 2000
1 + 2 = 3	10 + 20 = 30	100 + 200 = 300	1000 + 2000 = 3000
1 + 3 = 4	10 + 30 = 40	100 + 300 = 400	1000 + 3000 = 4000
1 + 4 = 5	10 + 40 = 50	100 + 400 = 500	1000 + 4000 = 5000
1 + 5 = 6	10 + 50 = 60	100 + 500 = 600	1000 + 5000 = 6000
1 + 6 = 7	10 + 60 = 70	100 + 600 = 700	1000 + 6000 = 7000
1 + 7 = 8	10 + 70 = 80	100 + 700 = 800	1000 + 7000 = 8000
1 + 8 = 9	10 + 80 = 90	100 + 800 = 900	1000 + 8000 = 9000
1 + 9 = 10	10 + 90 = 100	100 + 900 = 1000	1000 + 9000 = 10000
2 + 2 = 4	20 + 20 = 40	200 + 200 = 400	2000 + 2000 = 4000
2 + 3 = 5	20 + 30 = 50	200 + 300 = 500	2000 + 3000 = 5000
2 + 4 = 6	20 + 40 = 60	200 + 400 = 600	2000 + 4000 = 6000
2 + 5 = 7	20 + 50 = 70	200 + 500 = 700	2000 + 5000 = 7000
2 + 6 = 8	20 + 60 = 80	200 + 600 = 800	2000 + 6000 = 8000
2 + 7 = 9	20 + 70 = 90	200 + 700 = 900	2000 + 7000 = 9000
2 + 8 = 10	20 + 80 = 100	200 + 800 = 1000	2000 + 8000 = 10000
2 + 9 = 11	20 + 90 = 110	200 + 900 = 1100	2000 + 9000 = 11000
3 + 3 = 6	30 + 30 = 60	300 + 300 = 600	3000 + 3000 = 6000
3 + 4 = 7	30 + 40 = 70	300 + 400 = 700	3000 + 4000 = 7000
3 + 5 = 8	30 + 50 = 80	300 + 500 = 800	3000 + 5000 = 8000
3 + 6 = 9	30 + 60 = 90	300 + 600 = 900	3000 + 6000 = 9000
3 + 7 = 10	30 + 70 = 100	300 + 700 = 1000	3000 + 7000 = 10000
3 + 8 = 11	30 + 80 = 110	300 + 800 = 1100	3000 + 8000 = 11000
3 + 9 = 12	30 + 90 = 120	300 + 900 = 1200	3000 + 9000 = 12000
4 + 4 = 8	40 + 40 = 80	400 + 400 = 800	4000 + 4000 = 8000
4 + 5 = 9	40 + 50 = 90	400 + 500 = 900	4000 + 5000 = 9000
4 + 6 = 10	40 + 60 = 100	400 + 600 = 1000	4000 + 6000 = 10000
4 + 7 = 11	40 + 70 = 110	400 + 700 = 1100	4000 + 7000 = 11000
4 + 8 = 12	40 + 80 = 120	400 + 800 = 1200	4000 + 8000 = 12000
4 + 9 = 13	40 + 90 = 130	400 + 900 = 1300	4000 + 9000 = 13000
5 + 5 = 10	50 + 50 = 100	500 + 500 = 1000	5000 + 5000 = 10000
5 + 6 = 11	50 + 60 = 110	500 + 600 = 1100	5000 + 6000 = 11000
5 + 7 = 12	50 + 70 = 120	500 + 700 = 1200	5000 + 7000 = 12000
5 + 8 = 13	50 + 80 = 130	500 + 800 = 1300	5000 + 8000 = 13000
5 + 9 = 14	50 + 90 = 140	500 + 900 = 1400	5000 + 9000 = 14000
6 + 6 = 12	60 + 60 = 120	600 + 600 = 1200	6000 + 6000 = 12000
6 + 7 = 13	60 + 70 = 130	600 + 700 = 1300	6000 + 7000 = 13000
6 + 8 = 14	60 + 80 = 140	600 + 800 = 1400	6000 + 8000 = 14000
6 + 9 = 15	60 + 90 = 150	600 + 900 = 1500	6000 + 9000 = 15000
7 + 7 = 14	70 + 70 = 140	700 + 700 = 1400	7000 + 7000 = 14000
7 + 8 = 15	70 + 80 = 150	700 + 800 = 1500	7000 + 8000 = 15000
7 + 9 = 16	70 + 90 = 160	700 + 900 = 1600	7000 + 9000 = 16000
8 + 8 = 16	80 + 80 = 160	800 + 800 = 1600	8000 + 8000 = 16000
8 + 9 = 17	80 + 90 = 170	800 + 900 = 1700	8000 + 9000 = 17000
9 + 9 = 18	90 + 90 = 180	900 + 900 = 1800	9000 + 9000 = 18000

**Table B: Shirakatsi's tables of addition
shown in Hindu-Arabic numbers.**

3. This teaches the value of 6, that 6 is made of 1 and 5, or 2

and 4, or 3 and 3. This is a lasting discovery.

As I mentioned above, the first tablet teaches the child to count from 1 to 10. In the second tablet the sums are in the same sequence, and 11 follows 10. Examination of subsequent tablets will teach the child continuity in counting.

2.2.2. Second Column: the Addition of Tens

The second column shows addition of tens. The first tablet in column two helps the child to count in tens: 10, 20, 30, etc. They will connect counting in tens with the order of letters in the alphabet. Having the first and the second columns side by side will help the child to compare them, and find similarities and differences. The child who has already mastered the addition of single digit numbers and has already worked out a method to ease the addition of single digit numbers will use the same method to add tens.

Similarly, Table B can be effectively used to teach addition. Having the first and the second columns side by side helps the child to compare them, and to find similarities and differences. Having 2+3=5 and 20+30=50 side by side, the child notices the similarities (2, 3, 5), and will notice the difference, 0's. The child will learn that 20+30=50, meaning 2 tens and 3 tens, make 5 tens or 50. This addition builds up the knowledge presented in the first column and teaches the idea of the place value.

In adding 20 to 30, the present day student who is taught to perform addition using the "Addition Algorithm" will invariably, by rote, write 20 and 30 in a column, and will proceed to respond: "0+0=0, 2+3=5, therefore, the answer is 50."

2.2.3. Third and Fourth Columns: the Addition of Hundreds and Thousands

Next is the third column, the addition of hundreds. Building on the knowledge gained from the first and second columns, the child develops better knowledge of the place value and learns the addition of hundreds.

The fourth column teaches the pattern of addition of ones, tens, hundreds, and thousands. This presents the idea of progression and continuity, the idea of infinity.

Likewise, in working with thousands, to add 2000 to 3000 the present day student who is taught to perform addition using the "Addition Algorithm," again by rote will invariably write 2000 and 3000 in a column, and will proceed to respond: "0+0=0, 0+0=0, 0+0=0, 2+3=5, therefore, the answer is 5000." The student fails to see the connection between 2+3, 20+30, 200+300, and 2000+3000, which in turn prevents noticing progression and continuity.

GOHAR MARIKYAN

3. Conclusion

Initially, one may argue that this is a very slow pace in learning, and children nowadays are smarter than they were in the 7th century, that children in the 21st century have calculators and computers, and consequently they do not need to learn how to add. What if, accidentally, he/she pushes the wrong button, or types one more 0?

Strong foundation guarantees success in learning. Other than teaching the child how to add, this method teaches the child to understand the meaning of addition, and creates good understanding about the value of numbers. The method teaches thinking and encourages reasoning. From an early age this method teaches the child to separate issues, to analyze, to ask questions and to look for solutions. As a result logical thinking and reasoning become second nature for the child. In Shirakatsi's own words, he starts his teaching from its fundamental level, and presents the concepts in a most simplified manner for thorough understanding.

I have reviewed a number of elementary level textbooks where instead of teaching addition with a systematic, yes, slow pace, the system jumps into the addition of arbitrary pairs of numbers. It is the previously mentioned "Addition Algorithm," an unknown and hard to pronounce word for the child. It alienates the child and creates unnecessary distraction. Invariably, this method resorts to counting bunches of sticks haphazardly, that is, the child counts 5 sticks, then counts 3 sticks and puts them next to the 5 sticks, and counts all sticks again to come up to a total of 8. The child will not connect 5+3 and 3+5, but will count 3 sticks, then will count 5 sticks and will put them next to the 3 sticks, and then count all of the sticks again for a total of 8, getting confused all along. Boring? Yes! I have seen how some of my grown up students count, using fingers. Teaching addition through repetitive drills teaches to memorize that 5+3 is 8, 4+7 is 11, 3+5 is 8, but does not teach why it is so. This does not encourage the child to think, to understand, to look for solutions. Drills are very boring for the child.

There are "algorithms" for everything. Students associate the word algorithm with 'steps' that they have to memorize. This eliminates the necessity to understand, and learning math becomes exercising memorization of 'steps.' Furthermore, students have to memorize all possible "problem/solution algorithm" combinations. "Too much common sense is missing today from our educational system." states Dr. Joseph Casbarro, the author of several books on teaching[9]. Learning mathematics through memorization is not a long lasting learning. Mathematics builds up on previously gained knowledge, and a weak

---

[9] Casbarro 2003, p. XVII

foundation is detrimental to learning mathematics. All these distractions, boredom, and frustrations make learning math distasteful, hard, if not impossible. Despite my own encouragement in understanding, versus memorizing, my grown up students habitually try to break the solution into various 'steps,' which makes a short time memorization easier at the expense of understanding the problem. The question "Why do we do these steps?" has been replaced by the question "What are the 'steps'?" Furthermore, students do not use their common sense to check if the answer of the problem makes any sense at all. Once I assigned a problem to my students to find out how much gas was needed to drive one mile if with six gallons one could drive 150 miles. One student immediately responded, "900 gallons," applying multiplication instead of division.

About the Author

Gohar Marikyan is an Associate Professor and an advisor of Mathematics and Technology and the convener of Science, Mathematics, and Technology at the Metropolitan Center of Empire State College of State University of New York. She holds a Ph.D. degree in Mathematics (research Ph.D. in Mathematical Logic) and MS degree in Computer Science (research master's degree in Theory of Algorithms). Dr. Marikyan continues to actively pursue her research in mathematical logic as it applies to computer science. She also is interested in the history of math education as it applies to the development of teaching methodology. The results and findings of her research have been presented in domestic and international conferences and published in a number of scientific journals.

Dr. Marikyan's work with students is in two broad areas: Mathematics and Computer Science. In addition to these areas she also teaches business related math and computer courses. She conducts studies in groups, one-on-one, on distance, blended, and online. Concurrently, keeping up with advances in technology, Professor Marikyan continues to create new courses in mathematics and computer science for her students.

Professor Marikyan maintains an ongoing contact with her students after graduation and encourages them to freely seek her help whenever they need guidance or reassurance in their academic goals and careers.

Index

References

1. Abrahamyan, A., 1944, Anania Širakac'u Matenadrut'iwn [The Works of Ananias of Širak], Yerevan (in Armenian).

2. Ball, W. W. R., 1901, A Short Account of the History of Mathematics, New York: Macmillan.

3. Casbarro, J., 2003, Test Anxiety & What You Can Do About It: A Practical Guide for Teachers, Parents, and Kids, Port Chester, NY: Dude Publishing/ National Professional Resources, Inc.

4. Hacikyan, A. J., Basmajian, G., Franchuk, E. S., Ouzounian, N., 2002, The Heritage of Armenian Literature, Detroit: Wayne State University Press, vol. II.

5. Hewsen, Robert H., 1968, "Science in Seventh-Century Armenia: Ananias of Širak", in Isis, Vol. 59, No. 1, (Spring, 1968), pp. 32-45.

6. Rosenqvist, J. O., 2005, "Byzantine Trebizond: a Provincial Literary Landscape", in Byzantino-Nordica 2004. Papers presented at the international symposium of Byzantine studies held on 7-11 May 2004 in Tartu, Estonia, P. Ivo & J. Päll (eds.), Tartu, Estonia, Acta Societatis Morgensternianae, pp. 29-51.

7. Shirakatsi, A., 7th Century, Tvabanutiun (Arithmetic), Manuscript.

8. Shirakatsi, A., 7th Century, Autobiography, Manuscript.

9. Aghayan, E.B. (Editor), 1979, Anania Shirakatsi, Selected Works, Yerevan: Sovetakan Grokh.

Gohar Marikyan, Ph.D.

ANANIA SHIRAKATSI'S TVABANUTIUN: WORLD'S

OLDEST MANUSCRIPT ON ARITHMETIC

PART 1: ADDITION

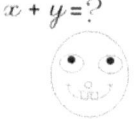

$x + y = ?$

New York, U.S.A., 2011

ISBN-13: 978-1463761783
ISBN-10: 1463761783